1 MONTH OF
FREE
READING

at
www.ForgottenBooks.com

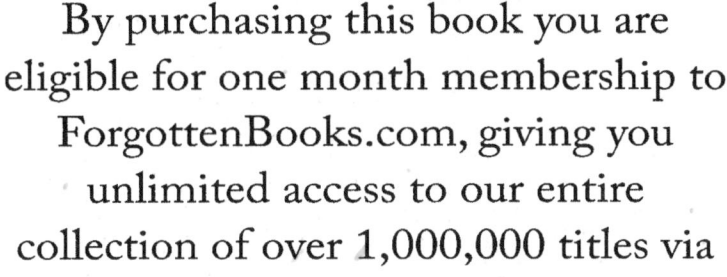

By purchasing this book you are eligible for one month membership to ForgottenBooks.com, giving you unlimited access to our entire collection of over 1,000,000 titles via our web site and mobile apps.

To claim your free month visit:
www.forgottenbooks.com/free903103

ISBN 978-0-265-87651-0
PIBN 10903103

VARIATION IN SEEDS AND SEEDLINGS

OF THE RADISH

BY

GIKAN FUJIMURA

B. S. University of Illinois, 1911

———————

THESIS

Submitted in Partial Fulfillment of the Requirements for the

Degree of

MASTER OF SCIENCE

IN BOTANY

IN

THE GRADUATE SCHOOL

OF THE

UNIVERSITY OF ILLINOIS

1911

ভ্রাম্বা
বিবেক

UNIVERSITY OF ILLINOIS

THE GRADUATE SCHOOL

May 31 1911   190

I HEREBY RECOMMEND THAT THE THESIS PREPARED UNDER MY SUPERVISION BY

Gikan Fujimura

ENTITLED   Variation In Seeds and Seedlings of the Radish

BE ACCEPTED AS FULFILLING THIS PART OF THE REQUIREMENTS FOR THE

DEGREE OF   Master of Science

Chas. F. Hottes
In Charge of Major Work

Burrill
Head of Department

Recommendation concurred in:

} Committee

on

Final Examination

CONTENTS

INTRODUCTION

Within any given species of plants as occurring in
nature, there appears a wide variation in form, in structure,
and in function. No one individual is the exact image of
another. A sufficiently large group or aggregation of individuals studied with reference to the variation of a single
character and expressed in the form of a curve, will agree
very nearly with another curve representing the variation of
the same character in a similar group or aggregation of organisms. If at the same time the variation of another character be studied in the identical groups of organisms, that
character may show a very different variability as expressed
in a curve, and yet the two, when properly studied, may be
shown to possess a definite interrelation or correlation.
That is to say, some characters tend, in greater or less
degree, to move together. This correlation of characters is
of the utmost importance and deserves the well merited at-
tention that all recent studies on variation have given it.

But, after all, the characters thus studied are, in
many cases, simply the external expression of functional
variation. In recent years, the interest in this form of
variation has greatly increased and for obvious reasons.

In the work that follows, it will be shown that a
high correlation exists between specific gravity of seeds and

power of growth of seedlings. The practical value of this information is obvious. In all samples of any seed there is a
great variation in the size and weight of the individual seeds.
Owing to their small size, it is extremely difficult to distinguish the seeds of high specific gravity from those of low
specific gravity. Size of seed does not always indicate the
specific gravity of a seed. Careful comparative tests have
proved that the best developed, most vigorous and uniform plants
are always produced from large seeds of high specific gravity;
the seeds of low specific gravity, irrespective of size, show
little uniformity in rate of germination, vigor of growth, and,
as a rule, produce small and undesirable plants.

From the above it is evident that a study that esta-
blishes a correlation between seed and vigor of growth of the
plant arising from the same, that is, functional activity, is
of great interest and importance from the view point of the
scientist and the agriculturalist.

A concrete illustration will show the extent to which
characters really move together.

Selection, in practice, means guiding the variation
in one or more characters of the organism by cutting off all
those individuals which are changing in undesirable directions
and reserving for reproduction only those which differ advan-
tageously from the average.

In Botany and in Zoology a vast amount of material
along these lines has been accumulated, and the origin of a
number of races and varieties is historically established.

The theoretical problems underlying this subject are complex, and we need more investigation, demonstration, illustration, and discussion.

Within the last two decades the study of variation and correlation has made gigantic strides, both along the theoretical and the practical branches.

In the beginning of the last century, Lamarck founded the theory of a common descent for all living beings. Half a century later, that is, in 1859, Darwin demonstrated that all the individuals of a given species differ from one another to a greater or less extent, and that these differences increase or lessen the chances of survival of the organism. Shortly after the publication of Darwin's Origin of Species, the Belgian anthropologist, Quetelet, submitted the variability in size and proportion of the different parts of the human body to a statistical investigation. He discovered, for example, that the variation in size of these parts follows a distinct law, and that this law agrees, in the main, with the law of probability. A. Weismann of Freiburg, E. Haeckel of Jena, W. Bateson of Cambridge, H. DeVrie of Amsterdam, G. Mendel of Brunn, H. Vernon of Oxford, F. Galton of Oxford, and E. B. Wilson of Columbia have, in their effort to explain the origin of variation, added much to our knowledge concerning variation in general. Along practical lines, the work of A. D. Shamel (40), V. A. Clark (7), J. W. Davel (14), H. Nilsson, etc., has given us much valuable material, and more especially along the lines considered in the present paper.

THE GERMINATION OF A COMPOSITE SAMPLE

The seeds used in the series of experiments that follows were obtained at one time from one of the well known seedmen, were thoroughly mixed, and a sample taken for the test of the normal germination. The seeds used for all experiments were taken from this original stock. The variety of radish used is the one known in the trade as "Icicle".

One hundred seeds taken at random from the above sample were placed on the several thicknesses of moist blotting paper in a flat glass jar known as the Koch moist chamber, and covered with more moist paper. Care was taken to allow for a circulation of air. These jars were kept in a warm green-house, the temperature during the experiment varying from 16° to 20° C.

The radish seeds were examined frequently, and it was perceived that a change, aside from the swelling, soon took place in at least a part of the seeds. The seeds were counted and thrown away as soon as sprouts appeared.

Seeds may fail to germinate from a variety of causes even when exposed to the proper degree of warmth, moisture and oxygen. They may be too old; they may not have been sufficiently mature when gathered; they may have become too dry; they may have been subjected to freezing before sufficiently dry; they may have been stored while damp and thus subjected to undue heating; or they may have been damaged by insects or fungi either before or after maturity. Further, the period during which the dormant seed is capable to respond by active growth to the external

conditions of warmth, moisture, etc., is limited. The time
varies from a few days to many years, and yet when this limit
is reached there may be no evidence, external or internal, to
indicate the loss of vitality. So this experiment has been done
to find the vitality of seeds before starting the other experiments.

The germination record of the composite sample is as
follows,-

TABLE I

Date	Number of germinated								
Oct. 26, 10 sowed	Oct. 27 1st day	Oct. 28 2nd day	Oct. 29 3rd day	Oct. 30 4th day	Oct. 31 5th day	Nov. 1 6th day	Nov. 2 7th day	Nov. 3 8th day	Nov. 4 9th day
	12	52	14	2	2	0	0	0	0

From the above figures, we learn that 82% of the
seeds of a composite sample germinate. The largest number (52%)
germinated on the second day, and none germinated after the
fifth day. The result may be expressed in a curve as shown on
Plate 1.

PLATE 1   6

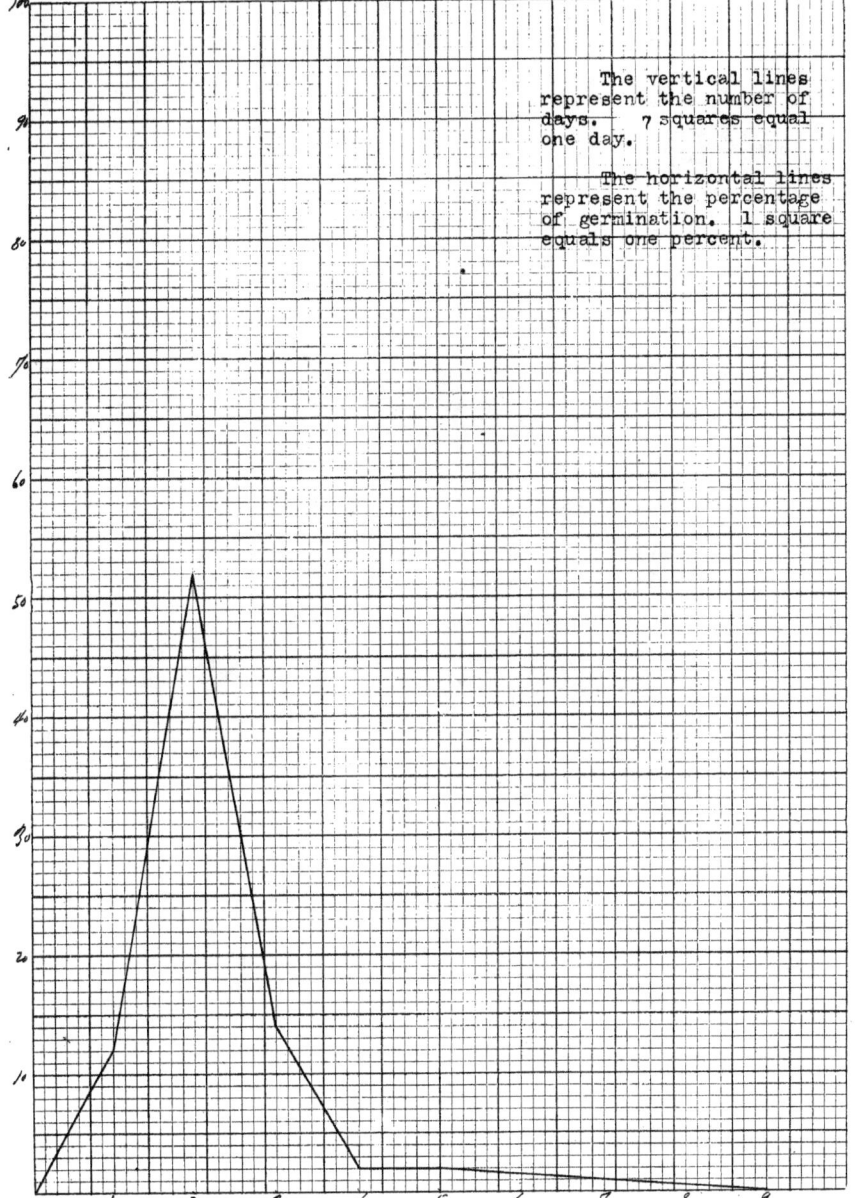

The vertical lines
represent the number of
days.   7 squares equal
one day.

The horizontal lines
represent the percentage
of germination. 1 square
equals one percent.

THE EFFECT OF LIGHT AND DARKNESS UPON THE GERMINATION AND RATE OF GROWTH OF A COMPOSITE SAMPLE OF RADISH SEEDS

As a rule, the best results are to be obtained by planting in sand in condition as nearly as possible approaching the normal requirement, and in all future experiments the seeds were sown in sand at a uniform depth. The process was somewhat as follows.

The germinators were shallow wood boxes 11 inches by 17 inches, inside measure, and were 3.5 inches in depth. Granite pans were also used. The pans or boxes were filled with moist sand and then tapped slightly to settle it, and, finally, struck off with a straight edge. A wire-net of 1/2 inch square mesh was placed on the surface of the sand. The seeds were sown very carefully at a uniform depth at equal distances. A record of the individual seeds was easily kept by the system of meshes of the wire-netting. The pans or boxes were prepared in duplicate and kept under similar conditions, except that one was exposed to direct light while the other was kept in darkness.

This experiment was carried on in a warm green-house, at 24° C. During the experiment care was taken that the sand was kept well moistened but never allowed to reach saturation. Germination in this series means the appearance of the seedling above the level of the sand.

The two sets of experiments, each of several hundred seeds, show that the rate and percentage of germination are

somewhat greater in the case of the seeds kept in darkness. The
tables below show in detail the results.

TABLE 2

Spe-cies	Date	Con-dition	Per cent of germination								
			4th day	5th day	6th day	7th day	8th day	10th day	11th day	12th day	13th day
Raph-anas Sati-vas	Nov. 25-10 Dec. 9-10	Light	6	26	45	57	65	68	72	73	73
	Nov. 28-10 Dec. 9-10	Dark	8	29	48	67	69	75	76	76	78

TABLE 3

Spe-cies	Date	Con-dition	828 seeds used in each case						
			4th day	5th day	6th day	7th day	8th day	9th day	10th day
Raph-anas Sati-vas	Dec. 19-10 Dec. 29-10	Light No. of seeds germ.	310	486		553	564	573	577
		Per cent	37	59		66	68	69	70
	Dec. 20-10 Dec. 30-10	Dark No. of seeds germ.	339		586	625	648	652	655
		Per cent	41		71	75	78	79	79

The tables show that the germination of seeds in the
darkness is 5% (Table 2) and 9% (Table 3) higher than in the
light.

The seedlings of the previous experiments were measured
daily for a period of 8 days after germination with the view of
determining vigor of growth, and fresh weight of leaves and
caulicles.

At the end of the period of observation, all the seedlings were pulled out from the sand, the leaves and caulicles
carefully removed with a very sharp knife and weighed immediately
so that they would not lose weight by drying.

The total height of the plants at the end of the 8 days
in the light is 2688 mm. as compared with a total height of 6072
mm. in darkness. The average height of the individual plant at
the end of that period is 37.1 mm. and 77.8 mm. The growth of
the leaves, however, as measured by their weight is 0.047 grms.
for light and 0.038 grms. for darkness. The caulicle of the
seedlings of a composite sample of radish seeds show a little
more than twice the total length in darkness than in light. On
the other hand, the leaves from the same seedlings show a difference of 0.009 grms. in favor of those grown in the light. This
striking feature is especially well shown in the photographs on
Plates 3, 4, and 5.

TABLE 4

Spe-cies	Date	Con-dition	Total height of plants								
			5th day	6th day	7th day	8th day	10th day	11th day	12th day	13th day	
	Nov. 25-10		1	22	46	56	66	71	72	72	No.of plants germ
Raph-anas	Dec. 9-10	Light10mm.	260	608	1218	2106	2312	2647	2688		Total hght of plants
Sati-vas		10mm.	11.8	13.2	21.7	31.9	32.5	36.7	37.1		Avge.hght.
	Nov. 25-10		1	15	59	66	71	72	75	78	No.of plants germ
	Dec. 9-10	Dark 9mm.	195	752	2101	4021	4435	5356	6072		Total hght of plants
		9mm.	13	12.7	31.8	56.6	61.5	71.4	77.8		Avge.hght.

TABLE 5

Condition	No.of seedlings		Total wt.	Av. weight
Light	68	stems	2.6 gr.	0.038 gr.
		leaves	3.2 gr.	0.047 gr.
Dark	72	stems		0.078 gr.
		leaves		0.038 gr.

Various explanations of the remarkable effect of the
absence of light in diminishing the size or preventing the
expansion of leaf blade, and the effect of light in retarding
the rate of growth in length have been offered; but it will not
be my object to consider this phase of the subject.

PLATE 2   11

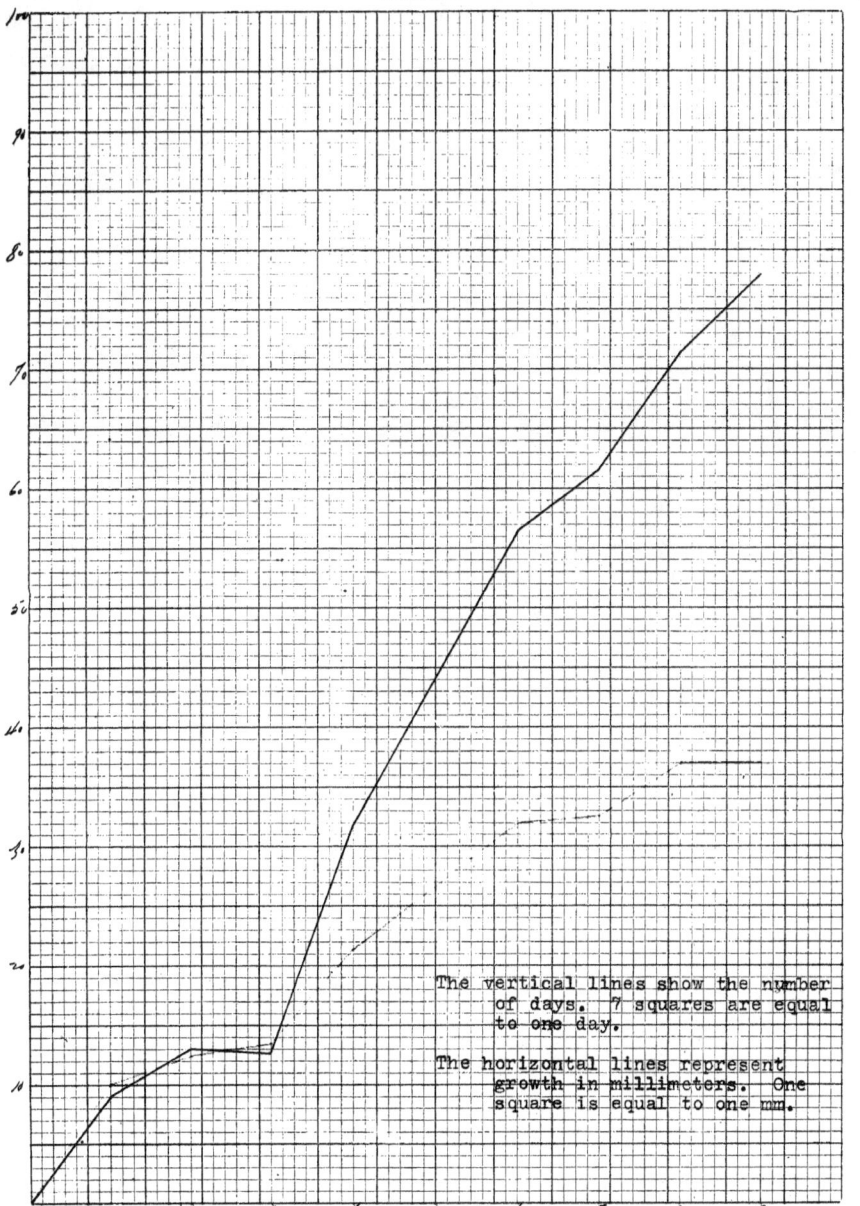

The vertical lines show the number
of days. 7 squares are equal
to one day.

The horizontal lines represent
growth in millimeters. One
square is equal to one mm.

PLATE 3

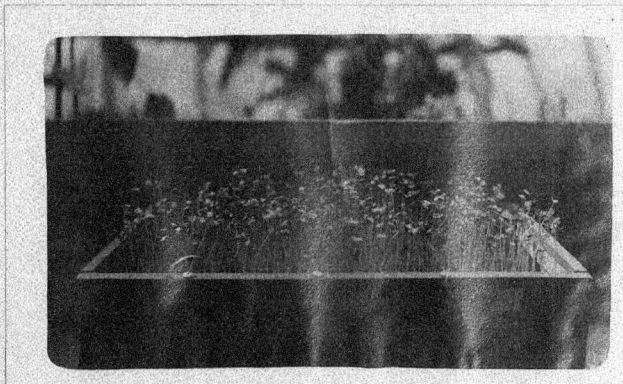

PLATE 4

PLATE 5

THE DISTRIBUTION OF THE SEEDLINGS ACCORDING TO RATE OF GROWTH

The seed from a composite sample was sown in sand in shallow boxes as already described. One box with 544 seedlings was placed in bright light, and another with 606 seedlings was placed under similar conditions except that the seedlings were kept in the dark. Ten days after sowing the seed, the seedlings were measured. The result is presented in tabular form just as it appears in the actual laboratory record. As a result of the development of statistical methods and their application to biological problems, we recently came to recognize the fact that this form is the most convenient and useful method in which to make the original record. A line stands for every individual examined.

The record of growth during a period of 8 days of 544 seedlings in light (Plate 6) shows 4 individuals of 10 mm. as the lowest and one individual of 110 mm. as the highest growth. Seventy-four show a growth of 70-75 mm. A curve plotted to indicate the frequency is shown on Plate 8.

The record of 606 seedlings grown in the dark and tabulated as shown on Plate 7, shows the extreme to be 15 mm. with a frequency 74, and 150 mm. with a frequency of 1. The mode has shifted from 70-75 in the light to 85-90 in the dark. A curve plotted to indicate the frequency is shown on Plate 8.

PLATE 6

CAULICLES IN THE LIGHT

Length of caulicles or value-v		No. of caulicles or frequency - F
10 mm.	////	4
15	//	2
20	///	3
25	////	5
30	//// ////	10
35	//// ////	10
40	//// //// //// //// ////	24
45	//// //// //// //// //// //// //	32
50	//// //// //// //// //// //// //// //// /	41
55	//// //// //// //// //// //// //// ////	40
60	//// //// //// //// //// //// //// //// //// ////	70
65	//// //// //// //// //// //// //// //// //// ////	53
70	//// //// //// //// //// //// //// //// //// ////	74
75	//// //// //// //// //// //// //// //// //// ///	48
80	//// //// //// //// //// //// //// //// //// ////	51
85	//// //// //// //// //// //// //// //// //	42
90	//// //// //// //// ///	23
95	//// //	7
100	///	3
105	/	1
110	/	1

PLATE 7
CAULICLES IN THE DARKNESS

Length of caulicles or value-V		No.of caulicles or frequency - F
15 mm.	////	4
20	////	4
25	/	1
30	////	4
35	////	5
40	//// //	7
45	//// //// ///	13
50	//// //// //	12
55	//// //// //// //// //	22
60	//// //// //// //// //// //// ////	34
65	//// //// ////	15
70	//// //// ////	15
75	//// //// //// //// //// //// ////	35
80	//// //// //// //// //// //// //// //// //// ////	50
85	//// //// //// //// //// //// //// //// //// //// //// //// //// //	67
90	//// //// //// //// //// //// //// //// //// //// //// //// //	62
95	//// //// //// //// //// //// //// //// //// //// //// //// /	61
100	//// //// //// //// //// //// //// //// //// //// //// //// /	61
105	//// //// //// //// //// //// //// //	37
110	//// //// //// //// //// //// //	32
115	//// //// //// //// ////	20
120	//// //// //// //// ////	25
125	//// /	8
130	//// ////	9
135	////	5
150	/	1

PLATE 8   18

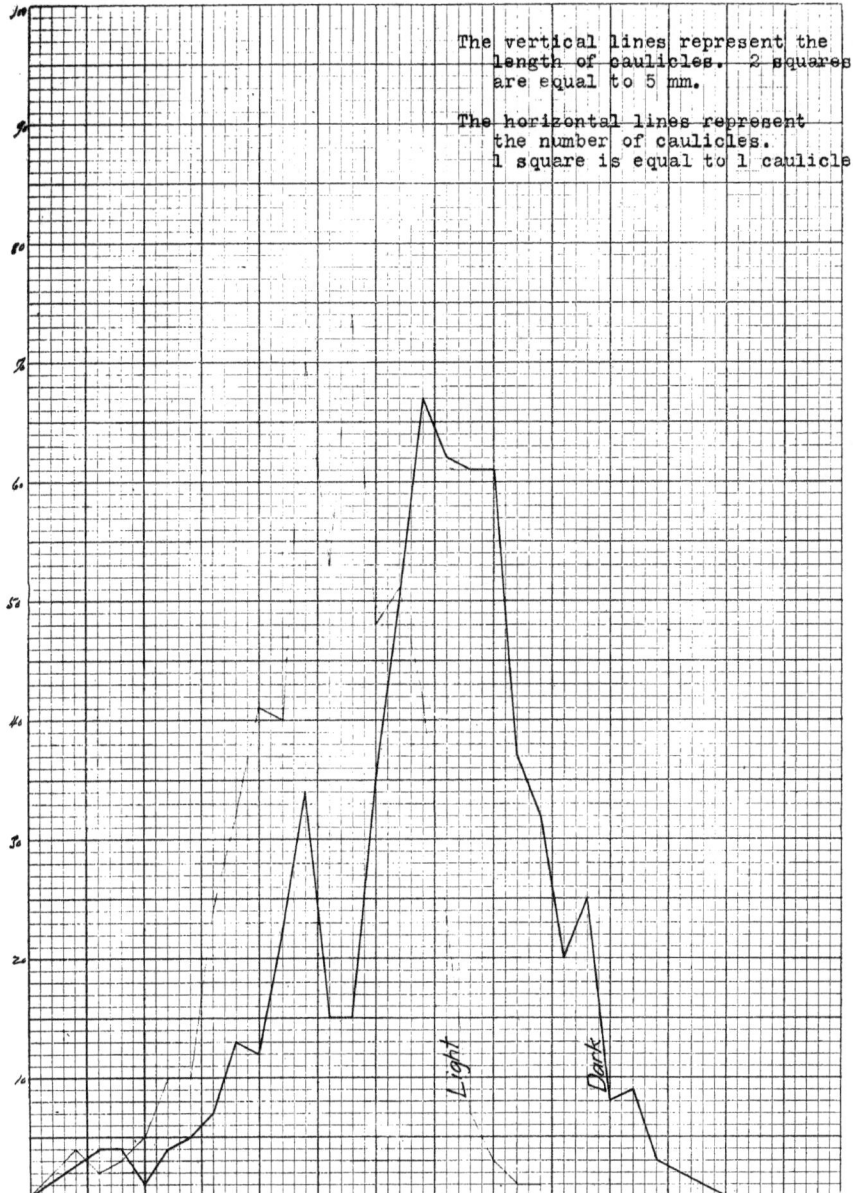

The vertical lines represent the
length of caulicles. 2 squares
are equal to 5 mm.

The horizontal lines represent
the number of caulicles.
1 square is equal to 1 caulicle.

THE DETERMINATION OF MEAN LENGTH OF CAULICLES
OF SEEDLINGS GROWN IN LIGHT AND IN DARKNESS

From data presented in the tables on Plates 6 and 7, the mean length of the caulicles was determined as follows: each value was multiplied by its frequency, the results added, and this sum divided by the number of individuals or variates. As shown on Plate 9 and 10, the mean length of the caulicle of seedlings grown in light is 64 mm., and of those in darkness 86 mm.

PLATE 9

V.		F.		FV.
1.0 cm.	x	4	=	4.0
1.5	x	2	=	3.0
2.0	x	3	=	6.0
2.5	x	5	=	12.5
3.0	x	10	=	30.0
3.5	x	10	=	35.0
4.0	x	24	=	96.0
4.5	x	32	=	144.0
5.0	x	41	=	205.0
5.5	x	40	=	220.0
6.0	x	70	=	420.0
6.5	x	53	=	344.5
7.0	x	74	=	518.0
7.5	x	48	=	360.0
8.0	x	51	=	408.0
8.5	x	42	=	358.0
9.0	x	23	=	207.0
9.5	x	7	=	66.5
10.0	x	3	=	30.0
10.5	x	1	=	10.5
11.0	x	1	=	11.0

544   3488 ÷ 544 = 64 - the mean length of
caulicle in mm.

21

PLATE 10

V.		F.		FV.
1.5 cm.	x	4	=	6.0
2.0	x	4	=	8.0
2.5	x	1	=	2.5
3.0	x	4	=	12.0
3.5	x	5	=	17.5
4.0	x	7	=	28.0
4.5	x	13	=	58.5
5.0	x	12	=	60.0
5.5	x	22	=	121.0
6.0	x	34	=	204.0
6.5	x	15	=	97.5
7.0	x	15	=	105.0
7.5	x	35	=	262.5
8.0	x	50	=	400.0
8.5	x	67	=	569.5
9.0	x	62	=	558.0
9.5	x	61	=	579.5
10.0	x	61	=	610.0
10.5	x	37	=	388.5
11.0	x	32	=	352.0
11.5	x	20	=	230.0
12.0	x	25	=	300.0
12.5	x	8	=	100.0
13.0	x	9	=	117.0
13.5	x	3	=	40.5
15.0	x	1	=	15.0
		606		5242.5

5242.5 ÷ 606 = 86 - the mean length of caulicle in mm.

22

THE DETERMINATION OF THE AVERAGE DEVIATION
OF THE CAULICLES OF SEEDLINGS GROWN IN LIGHT AND IN DARKNESS

Again using the data presented in the tables on Plates
9 and 10, the average deviation from the mean is shown on Plates
11 and 12. Column V equals variate; column F equals frequency;
column D equals deviation from mean 64 mm. Thus the first
caulicle deviates from the mean the difference between 10 mm.
and 64 mm., or 54 mm., and, being below the mean, it is written
with negative sign.

The results show that the average deviation of the
seedling grown in light is 13 mm., and of seedlings grown in
darkness 17 mm.

23

PLATE 11

V.	F.	D.	F.		D.		DF.
1.0om.	4	− 5.4	4	x	5.4	=	21.6
1.5	2	− 4.9	2	x	4.9	=	9.6
2.0	3	− 4.4	3	x	4.4	=	13.2
2.5	5	− 3.9	5	x	3.9	=	19.5
3.0	10	− 3.4	10	x	3.4	=	34.0
3.5	10	− 2.9	10	x	2.9	=	29.0
4.0	24	− 2.4	24	x	2.4	=	57.6
4.5	32	− 1.9	32	x	1.9	=	60.8
5.0	41	− 1.4	41	x	1.4	=	57.4
5.5	40	− 0.9	40	x	0.9	=	36.0
6.0	70	− 0.4	70	x	0.4	=	28.0
6.5	53	0.1	53	x	0.1	=	5.3
7.0	74	0.6	74	x	0.6	=	44.4
7.5	48	1.1	48	x	1.1	=	52.8
8.0	51	1.6	51	x	1.6	=	81.6
8.5	42	2.1	42	x	2.1	=	88.2
9.0	23	2.6	23	x	2.6	=	59.8
9.5	7	3.1	7	x	3.1	=	21.7
10.0	3	3.6	3	x	3.6	=	10.8
10.5	1	4.1	1	x	4.1	=	4.1
11.0	1	4.6	1	x	4.6	=	4.6
	544		544				740.2 ÷ 544 = 13mm

24

PLATE 12

V.	F.	D.	F.		D.		DF.
1.5 cm.	4	− 7.1	4	x	7.1	=	28.4
2.0	4	− 5.6	4	x	6.6	=	26.4
2.5	1	− 6.1	1	x	6.1	=	6.1
3.0	4	− 5.6	4	x	5.6	=	22.4
3.5	5	− 5.1	5	x	5.1	=	25.5
4.0	7	− 4.6	7	x	4.6	=	32.2
4.5	13	− 4.1	13	x	4.1	=	53.3
5.0	12	− 3.6	12	x	3.6	=	43.2
5.5	22	− 3.1	22	x	3.1	=	68.2
6.0	34	− 2.6	34	x	2.6	=	88.4
6.5	15	− 2.1	15	x	2.1	=	31.5
7.0	15	− 1.6	15	x	1.6	=	24.0
7.5	35	− 1.1	35	x	1.1	=	38.5
8.0	50	− 0.6	50	x	0.6	=	30.0
8.5	67	− 0.1	67	x	0.1	=	6.7
9.0	62	0.4	62	x	0.4	=	24.8
9.5	61	0.9	61	x	0.9	=	54.9
10.0	61	1.4	61	x	1.4	=	85.4
10.5	37	1.9	37	x	1.9	=	70.3
11.0	32	2.4	32	x	2.4	=	78.8
11.5	20	2.9	20	x	2.9	=	56.0
12.0	25	3.4	25	x	3.4	=	85.0
12.5	8	3.9	8	x	3.9	=	31.0
13.0	9	4.4	9	x	4.4	=	39.6
13.5	3	4.9	3	x	4.9	=	14.7
15.0	1	6.4	1	x	6.4	=	6.4
	606		606				6.9 ÷ 606 = 12 mm

25

THE CALCULATION OF THE STANDARD DEVIATION OF THE
CAULICLE OF SEEDLINGS GROWN IN LIGHT AND IN DARKNESS

The standard deviation may be determined from the data
given in the tables on Plates 11 and 12 as follows. The various
deviations are squared to eliminate the minus sign, and each
of these squared deviations is next multiplied by its respective
frequency. In the table that follows on Plates 13 and 14, V
equals variate; F equals frequency; D equals average deviation;
D^2F equals the squared deviation multiplied by its respective
frequency; Σ is the sign of summation.

Dividing the sum of D^2F (1557.53) by the frequency
(544) and extracting the square root of the quotient, we find
the standard deviation for the caulicle of seedlings grown in
light to be 17 mm. In a similar manner we find the standard deviation of the caulicle of seedlings grown in the dark to be
22.7 mm.

PLATE 13

V.	F.	D.	D^2	D^2F.
1.0 cm.	4	− 5.4	29.2	116.64
1.5	2	− 4.9	24.01	47.80
2.0	3	− 4.4	19.8	58.10
2.5	5	− 3.9	15.1	76.20
3.0	10	− 3.4	11.47	114.70
3.5	10	− 2.9	8.41	84.10
4.0	24	− 2.4	5.74	138.20
4.5	32	− 1.9	3.61	115.80
5.0	41	− 1.4	1.96	80.40
5.5	40	− 0.9	0.81	32.40
6.0	70	− 0.4	0.16	11.19
6.5	53	− 0.1	0.01	0.53
7.0	74	0.6	0.36	26.53
7.5	48	1.1	1.21	59.30
8.0	51	1.6	2.56	130.90
8.5	42	2.1	4.4	185.20
9.0	23	2.6	6.76	155.48
9.5	ı	3.1	9.61	67.20
10.0	3	3.6	12.96	38.90
10.5	1	4.1	16.81	16.81
11.0	1	4.6	21.15	21.15
	544			Σ 1557.53

Σ 1557.53 ÷ 544 = 2.9

$\sqrt{2.9} = 1.7$ cm.

PLATE 13

V.	F.	D.	D^2	$D^2F.$
1.0 cm.	4	- 5.4	29.2	116.64
1.5	2	- 4.9	24.01	47.80
2.0	3	- 4.4	19.8	58.10
2.5	5	- 3.9	15.1	76.20
3.0	10	- 3.4	11.47	114.70
3.5	10	- 2.9	8.41	84.10
4.0	24	- 2.4	5.74	138.20
4.5	32	- 1.9	3.61	115.80
5.0	41	- .4	1.96	80.40
5.5	40	- 0.9	0.81	32.40
6.0	70	- .4	0.16	11.19
6.5	53	- .1	0.01	0.53
7.0	74	.6	0.36	26.53
7.5	48	.1	1.21	59.30
8.0	51	.6	2.56	130.90
8.5	42	.1	4.4	185.20
9.0	23	2.6	6.76	155.48
9.5	7	3.1	9.61	67.20
10.0	3	3.6	12.96	38.90
10.5	1	4.1	16.81	16.81
11.0	1	4.6	21.15	21.15
	544			Σ 1557.53 ÷ 544 = 2.9

27

PLATE 14

V.	F.	D.	D^2.	D^2F.
1.5 cm.	4	− 7.1	50.41	201.64
2.0	4	− 6.6	43.50	174.24
2.5	1	− 6.1	37.20	37.21
3.0	4	− 5.6	31.36	125.44
3.5	5	− 5.1	26.01	130.05
4.0	7	− 4.6	21.16	148.12
4.5	13	− 4.1	16.81	218.53
5.0	12	− 3.6	12.96	155.52
5.5	22	− 3.1	9.61	211.42
6.0	34	− 2.6	6.76	229.84
6.5	15	− 2.1	4.41	66.15
7.0	15	− 1.6	2.56	38.40
7.5	35	− 1.1	1.21	42.35
8.0	50	− 0.6	0.36	18.00
8.5	67	− 0.1	0.01	0.67
9.0	62	0.4	0.16	9.92
9.5	61	0.9	0.81	43.41
10.0	61	1.4	1.96	119.56
10.5	37	1.9	3.61	133.51
11.0	32	2.4	5.76	184.32
11.5	20	2.9	8.41	168.20
12.0	25	3.4	11.56	289.00
12.5	8	3.9	15.21	121.68
13.0	9	4.4	19.36	174.24
13.5	3	4.9	24.01	72.03
15.0	1	6.4	40.96	40.96

$\overline{606}$

$\sqrt{5.195} = 2.27$ cm.

$\Sigma\ \overline{3154.41} \div 606 = 5.195$

28

RATE OF GROWTH OF INDIVIDUAL SEEDLINGS FROM A COMPOSITE SAMPLE

In a previous experiment, I noted that the rate of germination and growth of a composite sample of radish seeds varied widely. It is of interest to note the time of germination and rate of growth of each individual seed. Seventy-two seeds were planted in a manner already described, and by means of the wire-netting the germination and the rate of growth were easily recorded.

Growth, in general, is an increase in volume. In the scientific treatment of the subject, we must distinguish the several processes, that is, the formation of new cells, the unfolding or extension and enlargement of cells already present, and the differentiation of these cells. In the production of new cells and in the increase in size of the cells, there are certain chemical and physical phenomena which always accompany the morphological changes. Growth may be either diffused throughout the entire organism or local, forming a factor of differentiation.

Vegetable growth does not take place unless there is an available supply of organic matter capable of assimilation, access of free oxygen, and a sufficiently high temperature. Even under external conditions which are as nearly constant as possible growth is not quite uniform in its rate. This irregularity in the rate of growth of individuals under identical conditions is due to differences inherent in the seed, that is, the functional variation.

29

In the table that follows, the percentage of germination of 72 seeds is shown.

TABLE 6

Spe-cies	Date	Per cent of germination								(72 seeds were used)	
Rapha-nas	Dec. 16-10 Jan. 4-11	4th day	5th day	6th day	7th day	8th day	10th day	12th day	13th day	16th day	19th day
Sati-vas	No.of seeds germ.	21	34	40	47	51	55	55	56	57	57
	Per cent	29	47	56	65	71	76	76	78	79	79

PLATE 15

A record of the growth of the individual seedlings follows.

	4th day	5th day	6th day	7th day	8th day	10th day	12th day	13th day	16th day	19th day
1		Ger	10 mm	29	46	72	75	78	82	85
2			9	30	48	78	80	82	85	85
3		Ger	15	36	58	77	77	78	80	81
4									Ger	5
5	Ger	14	29	29	31	34	36	36	40	40
6		Ger	14	17	25	39	39	40	41	41
			13	26	45	76	85	90	95	95
8			12	31	47	65	65	70	72	72
9				Ger	15	23	24	25	30	30
10						Ger	Dead			
11	Ger	10	15	35	50	66	70	75	75	76
12		Ger	13	20	26	34	36	36	42	42
13					Ger	37	55	70	72	72
14		Ger	9	24	25	26	27	Dead		
15	Ger	10	18	27	36	47	50	55	55	55
16				17	32	51	55	60	62	62
17									Ger	Dead
18	Ger	---	11	20	22	30	30	35	37	37
19		Ger	13	26	39	55	55	57	60	60
20	Ger	---	25	37	50	70	72	80	80	84
21					25	52	70	85	90	90
22		Ger	15	35	51	65	70	77	80	80
23	Ger	---	13	17	25	39	42	46	47	47

PLATE 15 (continued)

#										
24	Ger		15	26	40	64	64	70	70	71
25	Ger	---	20	34	40	59	65	70	70	70
26			Ger	12	15	17	17	20	20	
27	Ger	---	21	44	56	72	77	84	86	86
28			Ger	20	41	46	50	50	54	
29	Ger	---	13	20	25	35	38	45	45	45
30			10	26	45	74	78	84	85	85
31	Ger	---	20	31	37	52	52	60	60	63
32	Ger	---	23	33	40	60	65	69	70	70
33				15	25	48	52	62	65	65
34	Ger	---	10	20	38	61	70	72	72	73
35	Ger	---	21	31	38	54	55	56	56	58
36						10	25	32	32	35
37	Ger	---	19	39	50	61	65	72	72	72
38					Ger	20	24	25	30	30
39	Ger	---	13	24	30	44	44	51	51	51
40	Ger	---	15	22	25	40	41	46	46	50
41					8	30	34	46	47	47
42		Ger	13	30	44	68	66	70	70	75
43		Ger	13	29	42	64	65	72	72	75
44		Ger	---	26	37	55	57	63	63	68
45						18	30	32	41	41
46			Ger	20	40	69	70	78	82	82
47	Ger	---	11	20	23	35	39	45	45	50
48			15	20	22	35	35	35	40	40

32

PLATE 15 (continued)

49	Ger	---	14	15	22	30	30	30	30	30
50	Ger	---	14	20	26	40	44	50	50	52
51						Ger	15	34	40	42
52	Ger	---	10	20	25	35	40	44	45	47
53			12	30	45	64	64	66	70	70
54				Ger	12	12	14	16	16	20
55	Ger	---	16	24	35	45	46	50	54	54
56		Ger	14	30	45	65	70	75	75	75
57		Ger	17	32	45	65	66	75	75	76

Table 7 is a tabulation of the above records in a manner that will show the correlation between time of germination and rate of growth.

An analysis of this record shows little or no correlation between the time of germination and rate of growth. To be sure, the number of seedlings is rather too small to draw definite conclusions. The record does give, however, some indication of what may be effected when larger numbers are used.

TABLE 7

Length in mm.	4th day	5th day	6th day	7th day	8th day	10th day	12th day	13th day	16th day	19th day	Total
20				2							2
30	2			1	1	1					5
40	4	2	1		1	2					10
50	7			1							8
60	1	2		2							5
70	5	5	2		1						13
80	2	3	3								8
90			1		1						2
Total	21	12	7	6	4	3					53

34

THE RELATION OF SPECIFIC GRAVITY TO GERMINATION OF RADISH SEEDS

The most satisfactory means of separating the seeds
of low from those of high specific gravity is by the use of a
solution of $NaNO_3$.

The density of the solution was obtained in a manner
somewhat as follows. A saturated solution of $NaNO_3$ was tested
with Twaddell's hydrometer and its temperature and density
noted. The specific gravity of such a solution was found to
be 1.20. A definite quantity of this solution was then diluted
with distilled water to make solutions of specific gravity 1.05,
1.10, and 1.15. These four solutions, together with distilled
water, were then used as later described to determine the specific gravity of the seeds.

Twaddell's hydrometer is a direct reading instrument.
It is, however, not adapted to liquids lighter than water. The
system consists of a series of spindles (usually six in number)
carrying graduations from 0 to 174. The reading in pure water,
at 15.5° C. is taken as 0, and each subsequent rise of 0.005
specific gravity is recorded on the scale as one additional
division. Twaddell's readings are converted into specific gravity
as follows.

The reading on the spindle is multiplied by 0.005 and
1.000 is added to the product. Thus 15 Twaddell becomes 1.075
specific gravity,-- 15 x 0.005+ 1.000 = 1.075.

The seeds were placed in a large strainer which was
then dipped into alcohol of 95% for a few minutes to remove the

film of air adhering to the surface of the seeds and also to
dissolve or dislodge the dust.

The alcohol was thoroughly drained from the seeds
and the strainer with the seeds was now plunged into distilled
water at 15° C. The water was frequently changed and tested to
avoid error. All seeds of less specific gravity than 1.00
floated and were skimmed from the surface with a smaller strainer.
They were dried on filter paper and stored for future experiments. The strainer now containing seeds of specific gravities
greater than 1. was, after thorough draining, plunged into $NaNO_3$
solution of specific gravity 1.05. Again all seeds floating
were skimmed off the surface, rinsed in water, and placed on
filter paper to dry. These were labelled, seeds lighter than
specific gravity 1.05. In this manner the seeds from a composite
sample were separated as to specific gravity.

Hydrometer tests were made of the solution each time
before use and any error due to dilution corrected.

In testing for the relation between specific gravity
and germination, the boxes with sand already described were used.
One hundred seeds of each kind were carefully planted and the
sand sifted loosely over them. This method gave the very best
results.

36

Tables showing the germination of radish seeds
of different specific gravities.

TABLE 8

SpeciesDate	Sp. Gr.	4days	5days	6days	7days	8days	9days	10days	Total
Rapha- Jan.	Seeds lighter than the dis-								
nas 21	tilled water	28	36	38	40		40	40	
'11	Seeds lighter than Sp.Gr.1.05	40	46	47	47		49	50	
Sati- Jan.	Seeds lighter								
31	than Sp.Gr.1.10	46	56	62	64		66	66	
vas '11	Seeds lighter than Sp.Gr.1.15	54	58	59	61		62	62	
#	Seeds lighter than Sp.Gr.1.20	25	30	32	33		34	35	253

TABLE 9

SpeciesDate	Sp. Gr.	4days	5days	6days	7days	8days	9days	10days	Total
Raph- Jan.	Seeds lighter than the dis-								
anas 28	tilled water	8	31	46	49		52 ·	53	
'11	Seeds lighter than SpGr. 1.05	10	45	57	60		63	63	
Sati- Feb.	Seeds lighter								
7	than SpGr.1.10	13	54	65	68		71	72	
vas '11	Seeds lighter than SpGr. 1.15	28	61	66	70		77	77	265

In the case of specific gravity 1.20 only 98 seeds
were used, there being very few seeds of so high a specific
gravity.

37

TABLE 10

Spe-cies	Date	Sp. Gr.	Per cent of germination							
			4days	5days	6days	7days	8days	9days	10days	Total
Raph-anas	Feb. 3 '11	Seeds lighter than the dis-tilled water	28	45	51	53	54		55	
		Seeds lighter than SpGr. 1.05	24	33	44	51	52		56	
Sati-vas	Feb. 13 '11	Seeds lighter than SpGr. 1.10	62	71	75	78	78		79	
		Seeds lighter than SpGr. 1.15	53	73	80	86	87		88	278

The results of these three tests may for convenience be brought together in a table as shown below.

TABLE 11

Spe-cies	Specific Gravity	Per cent of germination						
		4days	5days	6days	7days	8days	9days	10days
Raph-anas	Seeds lighter than distilled water	21	37	45	47		49	49
	Seeds lighter than Sp.Gr. 1.05	35	41	49	53		55	56
Sati-vas	Seeds lighter than Sp. Gr. 1.10	40	60	67	70		72	72
	Seeds lighter than Sp.Gr. 1.15	45	64	68	72		75	76
	Seeds lighter than Sp.Gr. 1.20 #	25	30	32	33		34 '	35

From the above it is very apparent that the rate and percentage of germination increases with the specific gravity of the seed until it reaches specific gravity 1.20. Seeds of this specific gravity show a very marked decrease.

Curves plotted to show the relation between specific gravity and germination are shown on Plate 16.

PLATE 16   38

Lighter than distilled water

Lighter than Sp. Gr. 1.05

Lighter than Sp. Gr. 1.10

Lighter than Sp. Gr. 1.15

Lighter than Sp. Gr. 1.20

UNIVERSITY OF DENVER —S.C.A. FORM 1

PLATE 17

PLATE 18

THE RELATION BETWEEN SPECIFIC
GRAVITY AND WEIGHT OF SEEDS OF RAPHANAS SATIVAS

As shown on the table, one hundred seeds of each specific gravity were placed on a small watch glass and weighed on a chemical balance. Ten different lots were weighed. In each case the seeds were taken without selection. The results are shown on Plates 19 and 20.

PLATE 19

WEIGHT OF RAPHANAS SATIVAS   FEB. 14 - 16, 1911

No. of Exp.	100 seeds lighter than distilled water	100 seeds lighter than Sp. Gr. 1.05	100 seeds lighter than Sp. Gr. 1.10	100 seeds lighter than Sp. Gr. 1.15
1	0.795gr	0.832gr	0.852gr	0.945gr
2	0.793	0.851	0.961	0.820
3	0.770	0.816	0.834	0.860
4	0.788	0.890	0.882	0.924
5	0.746	0.800	0.895	0.854
6	0.755	0.817	0.800	0.825
	0.843	0.770	0.898	0.828
8	0.801	0.845	0.917	0.838
9	0.757	0.830	0.877	0.834
10	0.850	0.784	0.875	0.864
Average	0.790	0.824	0.879	0.858

97 J 5 c

The ten hundred seeds of specific gravity lighter than water, 1.05, 1.10, and 1.15 respectively, show an average weight of 0.790 grams, 0.824 grams, 0.879 grams, and 0.858 respectively. It is interesting to note that in the case of seeds lighter than distilled water, and specific gravity 1.05 and 1.10 there is a corresponding increase in weight per 100 seeds. In the case of specific gravity 1.20 the ten hundred seeds taken at random as 10 different samples show an average weight of 0.858 or .21 grams less than an equal number of seeds of specific gravity 1.15. From the tables it is apparent that a correlation between specific gravity and weight exists only in the case of the lower specific gravity. The cause for the decrease in weight of the samples of seeds of specific gravity 1.20 will be discussed in connection with the next experiment.

44

THE RELATION BETWEEN SPECIFIC GRAVITY AND SIZE

OF THE SEEDS OF RAPHANAS SATIVAS, AND

THE RELATION BETWEEN SIZE

AND THE GERMINATION

OF THE SEEDS

There has been a great diversity of opinion as to whether the vigor of a radish plant is or is not governed by the weight or the size of the seed from which it grew. This is an important question. If the lighter seeds or those of small size produce less vigorous plants, we should of course elimi- nate them from the seed radish used for sowing. It is the strongest, hardiest plants that are desired, and we are inter- ested in the seeds that give us the largest and best returns.

The seeds of specific gravity lighter than water and specific gravity 1.05, 1.10, 1.15, 1.20 respectively were in turn placed in sieves having perforations of different sizes and thus the seed of each specific gravity was divided accord- ing to size into small, medium, and large. One hundred small, medium, and large seeds of each specific gravity were weighed with the result given in the table below.

TABLE 12

Specific Gravity	Small	Medium	Large
Seeds lighter than water	0.448	0.850	1.116
Seeds lighter than Sp. Gr. 1.05	0.456	0.859	1.128
Seeds lighter than Sp. Gr. 1.10	0.538	0.884	1.245
Seeds lighter than Sp. Gr. 1.15	0.529	0.931	1.202

45

Again we note that the weight of the small, medium, and large seeds of the next higher specific gravity show a proportionate increase over the lower until we come to specific gravity 1.15. Here, as in the experiment previously described, we find the small and the large seeds show a distinct decrease in weight over the next lower. In the case of the medium seed of that specific gravity, a regular increase in weight is noted.

The germination of 100 small, medium, and large seeds of various specific gravity of Raphanas Sativas, planted Feb.25-11

TABLE 13

Grades of seed		Per cent of germination						
		4th day	5th day	6th day	7th day	8th day	9th day	10th day
Seeds lighter than water	Small	20	27	28	28		29	31
	Medium	37	38	41	44		44	44
	Large	29	33	34	34		34	36
Seeds lighter than S.G. 1.05	Small	10	15	17	19		20	20
	Medium	10	13	13	14		17	17
	Large	9	17	19	20		21	22
Seeds lighter than S.G. 1.10	Small	31	40	41	44		44	45
	medium	42	47	52	52		52	52
	Large	31	44	47	48		49	51
Seeds lighter than S.G. 1.15	Small	21	30	32	33		48	41
	Medium	28	39	46	50		53	53
	Large	42	49	52	56		59	59

46

From the above table it is apparent that, as a rule, the medium-sized seed in each case except one, shows the highest percent of germination and earliest germination. The small seeds are distinctively lower in rate and percent of germination than the large. These tables illustrate very plainly the existence of a correlation between specific gravity and size of seed. There is a gradual rise in percent and rate of germination from the small seeds to the large seeds, and from the seeds of lowest specific gravity to the seeds of highest specific gravity, except that in the case of specific gravity 1.15 the curve again falls.

47

THE RELATION BETWEEN SIZE OF SEED AND RATE OF GROWTH OF SEEDLINGS

In selecting seeds for planting they should be fully matured, of moderate size and perfect development; that is, they should be sufficiently developed to give them their full allotment of vigor from the mother plant from which they have been selected. The continued use of small seed gives weak plants, and this conclusion has been reached by a large number of investigators, both in this country and in Europe. The discarding of small seed is regarded as an important point/in the production of a maximum crop.

The seedlings studied were the same as used for the germination test just described. The caulicles were measured every day for six days after having reached 5 mm. in length. The final measurement was made 9 days after the sowing of the seed, and in no case was a seedling measured which was less than 3 days after the first measurement.

48

Table showing the relation between size and

specific gravity and the growth of caulicles

TABLE 14

Specific Gravity and sizes of seeds	Seeds used	Seeds germinated	Plants used for calculation	Average length of caulicles
Seeds lighter than water				
Small	100	31	28	35 mm.
Medium	100	44	38	47
Large	100	36	33	46
Seeds lighter than Sp. Gr. 1.05				
Small	100	20	16	28
Medium	100	17	13	37
Large	100	22	15	31
Seeds lighter than Sp. Gr. 1.10				
Small	100	45	41	34
Medium	100	52	52	37
Large	100	51	47	36
Seeds lighter than Sp. Gr. 1.15				
Small	100	41	31	33
Medium	100	53	45	35
Large	100	59	52	38

An examination and comparison of the data presented
in table 14, and also the curve and photograph, show that the
medium sized seed of each specific gravity grows most rapidly.

49

In the case of seeds of specific gravity less than distilled water we find the medium and large sized seeds possessed with a vigor of growth approximately 21% and 17% respectively greater than the seed of highest specific gravity.

PLATE 21

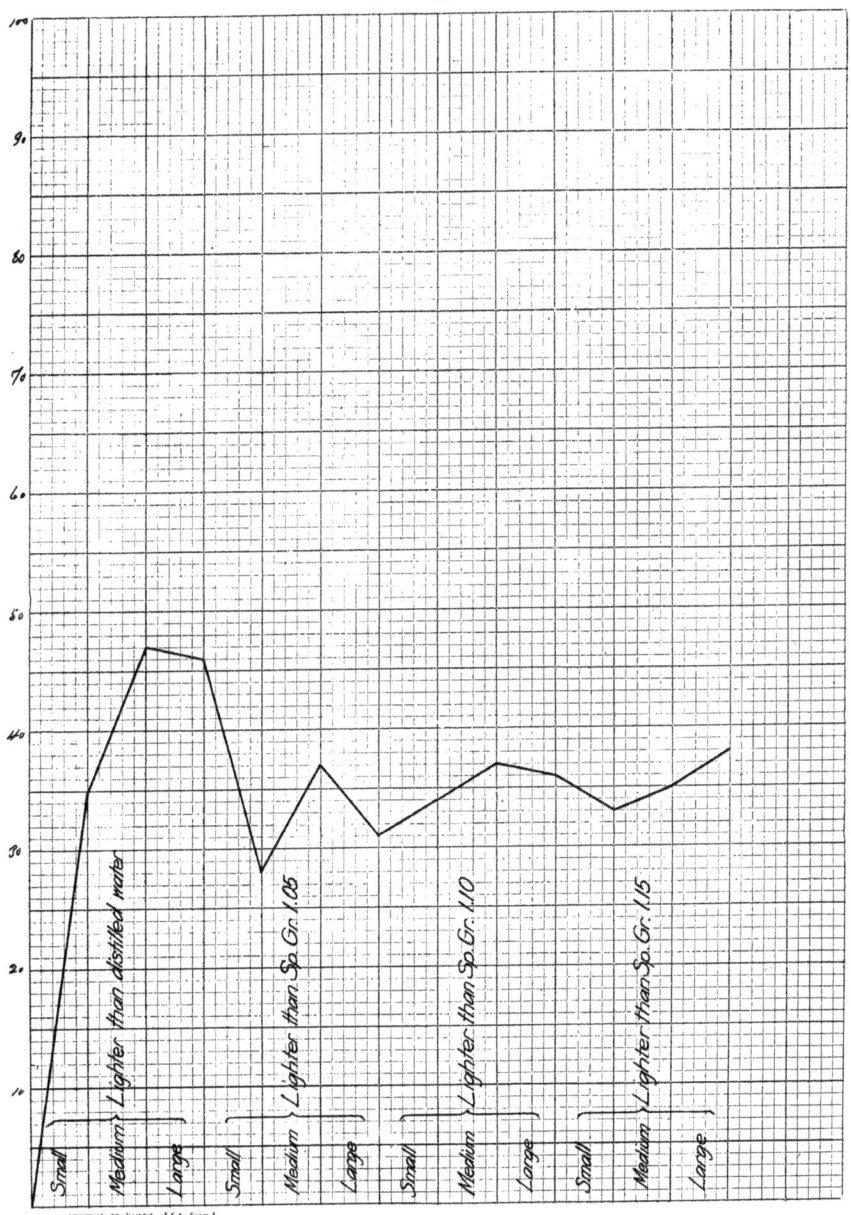

PLATE 21

Sp. Gr. 1.05

ghter than Sp. Gr. 1.10

Lighter than Sp. Gr. 1.15

Large

Small

Medium

Large

PLATE 22

THE CORRELATION BETWEEN SPECIFIC GRAVITY AND LENGTH OF CAULICLES

Any peculiarity in the development of one organ, or
set of organs, is usually accompanied by a corresponding modi-
fication or suppression of an organ belonging to some other
parts of the system, and this relation of different parts to
each other and to the whole is known as correlation. Correlation exists in all the stages of the development of an organism,
sometimes in one way, sometimes in another.

Vochting writes,- in a tree that is growing under
normal conditions, without being subjected to injury, all the
organs appear in definite relations to each other; so many
leaves correspond to a definite number of twigs and branches.
These spring from a stem of proportionate thickness, and the
stem passes into a definitely proportionated tap-root, from
which arise a due array of lateral roots. In normal conditions
all these organs are in equilibrium. An apple-tree, growing
on the lines where tilled garden ground meets a lawn, grows more
vigorously on the side toward the garden. If one of the roots of
an apple-tree with three main roots and three branches be com-
putated, then the corresponding branch will lag behind in growth,
although it may not absolutely perish. The equilibrium varies
according to the specific nature of the tree.

Again Goebel states that, the fact that lateral buds
do not develop while the axial bud is still growing vigorously
is because of the relation between the two. Vochting's experiments

upon beet-root are still more characteristic. "The stem of a
beet plant that bore young buds gave rise to vegetative shoots
when it was united with a young, still growing root, but to a
blossoming stem when it was grafted in spring, upon an old root."

   In all samples of radish seed there is great variation
in the weight of the individual seeds. Differences in specific
gravity are due either to differences in structure or in composition and the latter has a great influence upon the specific
gravity. The specific gravity of a seed is truly related to
the vigor of the resulting plant; this fact must presumably be
due to differences in composition of the seed, and these differences, moreover, of reserve material present.

   To determine the degree of correlation between the
length of caulicle and specific gravity of radish seeds, a so-
called "Correlation table" is constructed out of the measurements
of the two characters as found in a large number of individuals.
Knowing this relationship and the value of one of the characters,
we are enabled to calculate the corresponding mean value of the
other. The purpose of this experiment is to represent the results of an attempt to measure the intensity of the interrela-
tionship between the length of caulicle and specific gravity and
the advantages of this for purposes of selection are obvious.
The correlation table follows.

54

TABLE 15

	Length of caulicles in mm.								
	10	15	20	25	30	35	40	45	50
Lighter than distilled water	2	--	6	1	1	1	--	1	--
Lighter than Sp. Gr. 1.05	4	--	--	1	1	--	1	2	1
Lighter than Sp. Gr. 1.10	1	1	3	2	3	3	3	4	2
Lighter than Sp. Gr. 1.15	--	1	2	1	1	--	--	--	--
Lighter than Sp. Gr. 1.20	--	2	1	1	2	1	--	--	3
	7	4	12	6	8	5	4	7	6

TABLE 15 (continued)

	55	60	65	70	75	80	85	90	95	100	105
Water	5	3	4	3	2	1	1	2	1	3	1
SpGr. 1.05	1	3	--	5	7	3	4	2	4	1	1
SpGr. 1.10	2	--	5	3	5	3	5	5	--	6	1
SpGr. 1.15	2	2	2	3	9	11	4	1	6	3	3
SpGr. 1.20	3	1	2	3	5	4	1	--	1	1	1
	13	9	13	17	27	22	15	10	12	14	7

TABLE 15 (continued)

	110	115	120	125	130	135	
Water	1	1	--	--	--	--	40
SpGr. 1.05	2	1	2	2	1	--	49
SpGr. 1.10	2	1	2	--	1	1	64
SpGr. 1.15	5	3	--	--	1	--	60
SpGr. 1.20	--	1	--	--	--	--	33
	10	7	4	2	3	1	246

55

TABLE 16

	Length of caulicles in mm.							
	10	15	20	25	30	35	40	45
Lighter than distilled water	1	--	1	1	1	4	1	6
Lighter than Sp. Gr. 1.05	2	1	1	3	--	5	2	--
Lighter than Sp. Gr. 1.10	3	--	1	--	3	1	2	1
Lighter than Sp. Gr. 1.15	2	2	2	--	--	3	2	3
	8	3	5	4	4	13	7	10

TABLE 16 (continued)

	50	55	60	65	70	75	80	85
Water	4	6	4	4	5	4	5	4
SpGr. 1.05	4	6	5	4	6	8	1	5
SpGr. 1.10	2	10	5	4	11	3	10	5
SpGr. 1.15	5	6	9	5	7	5	8	5
	15	28	23	17	29	20	24	19

TABLE 16 (continued)

	90	95	100	105	110	115	
Water	--	1	--	--	1	--	53
SpGr. 1.05	2	3	2	1	1	1	63
SpGr. 1.10	5	1	4	--	--	1	72
SpGr. 1.15	9	--	2	1	--	--	76
	16	5	8	2	2	2	264

56

TABLE 17

	Length of caulicles in mm.							
	10	15	20	25	30	35	40	45
Lighter than distilled water	1	--	--	1	1	1	3	4
Lighter than Sp. Gr. 1.05	5	--	2	--	2	--	2	1
Lighter than Sp. Gr. 1.10	2	1	1	1	1	--	2	2
Lighter than Sp. Gr. 1.15	--	1	3	2	3	1	1	5
	8	2	6	4	7	2	8	12
Average of 3 exp.	8	3	8	5	6	7	6	10

TABLE 17 (continued)

	50	55	60	65	70	75	80	85	90
Water	5	2	7	6	3	5	8	1	5
Sp. Gr. 1.05	3	6	3	6	2	2	7	3	3
Sp. Gr. 1.10	5	2	6	7	9	11	17	6	3
Sp. Gr. 1.15	7	11	12	4	15	5	6	7	4
	20	21	28	23	29	23	38	17	15
Average of 3 exp.	14	21	20	18	25	23	28	17	14

TABLE 17 (continued)

	95	100	105		Average of 3 exp.
Water	1	--	--	54	49
Sp. Gr. 1.05	2	1	--	50	54
Sp. Gr. 1.10	1	--	1	78	71
Sp. Gr. 1.15	--	1	--	88	75
	4	2	1	270	
Average of 3 exp.	7	8	3		

57

The above data may be considered as representative
of average results, and from this we learn that the mean length
of caulicle is 80 mm, and there is a gradual increase in number
of seedlings from the seeds lighter than distilled water.
Careful comparative tests of light and heavy seeds have proved
that the best developed and most vigorous plants are always
produced from large and heavy seeds. The plants from the heavy
seeds grew more rapidly than those from the light seeds. The
heavy seeds also produced more uniform plants than the light
seeds, and the plant from the heavy seed produced greener and
wider leaves and were in general more vigorous.

58

VARIATION

Variation is the rule rather than exception in all organisms; therefore, a study of living things is not complete without a knowledge of their diversities. With the application of mathematics to the field of biology, studies along this line have assumed greater importance, for by the use of statistical methods we are able to determine with exactness the range of variation in the species as shown in the previous experiments, for a simple variety of radish.

The fluctuations from time to time are the result of environmental influence and little or unknown internal causes.

Changes in varieties are made by taking advantage of certain laws which govern all living things, both plants and animals. The first is the law of heredity, that "like begets like". Radish produces seed which brings forth radish, never by any chance wheat or corn. It is just here that another law comes in, a law which makes progress possible. This is the law of variation, the tendency of offspring to be unlike the parents. If the plant were in all respects exactly like its parents, no improvement would be possible.

Our plants, in common with other living forms, are endowed with a flexibility or plasticity in a large number of the organizations that enables them to adapt themselves to conditions the environment presents. Again there are changes that occur in so far as now present knowledge goes, independent of the environment.

59

Four distinctly different kinds of variations are recognized, that is morphological, substantive, meristic, and functional variations. Variation is both quantitative and qualitative, both continuous and discontinuous, and these distinctions should be clearly in mind at all times.

The principal causes of general variation in plants are numerous, but chiefly they can be classified into two, that is, ~~the~~ internal and external to the organism.

60

CONCLUSION

1. A test for determining power of germination, etc. of a composite sample of seeds to be of value must be made of a large number of seeds.

2. The composite sample of seeds of the radish used in these experiments shows that 82% germinated and that the highest percentage, 52%, was on the second day.

3. The average height of seedlings of the radish grown in light is 37.1 mm.; of those grown in darkness is 77.8mm.

4. The record of growth during a period of 8 days of 544 seedlings in light shows 4 individuals of 10 mm. as the lowest and one individual of 110 mm. as the highest growth. Seventy-four show a growth of 70 - 75 mm. The record of 606 seedlings grown in the dark and tabulated as shown on Plate 7, shows the extreme to be 15 mm. with a frequency 4, and 150 mm. with a frequency 1. The mode has shifted from 70-75 in the light to 85-90 in the dark.

5. The mean length of the caulicle of seedlings grown in light is 64 mm. and of those in darkness 86 mm.

6. The results show that the average deviation of the seedlings grown in light is 13 mm., and of seedlings grown in darkness 17 mm.

7. The standard deviation for the caulicle of seedlings grown in light is 17 mm.; of caulicles grown in darkness is 22.7 mm.

8. From the number of seeds used, we must conclude

that there is little or no correlation between time of germination
and rate of growth.

9. The rate and percentage of germination increase
with the specific gravity of the seed until it reaches specific
gravity 1.20. Seeds of this specific gravity show a very marked
decrease.

10. A correlation between specific gravity and weight
exists only in the case of the lower specific gravity.

11. The medium-sized seed, irrespective of specific
gravity, shows the highest rate of germination.

12. The vigor of growth in each of the four specific
gravities tested was greatest in the medium-sized seed.

13. The best developed and most vigorous plants are
always produced from large and heavy seeds. The plants from
the heavy seed grew more rapidly than those from the light seeds.
The heavy seeds also produced more uniform plants than the light
seeds, and the plant from the heavy seed produced greener and
wider leaves, and were in general more vigorous.

62

BIBLIOGRAPHY

1. Arther, J.C., Delayed germination in the cocklebur and
      other paired seed, Proc.Soc.Prom.Agri.Sci. 16:70-79,
      1895.

2. Barnes, C.R., The effect of light, Outlines of plant life.

3. Bolley, H.L., Seed wheat, North Dakota Ag. Exp. St. Press
      Bul. 2.

4. Burwash, L.I., Seed selection according to specific gravity.

5. Carter, L.E., Vitality, adulteration, and impurity of
      clover, Iowa Ag. Exp. Sta. Bul. 88.

6. Clark, C.F., Variation and correlation in timothy, Cor-
      nell Univ. Ag. Exp. Sta. Bul. 279.

7. Clark, V.A., Seed selection according to specific gravity,
      N.Y. Ag. Exp. Sta. Bul. 256, October, 1904.

8. Crocker, W., Role of seed coats in delayed germination,
      Bot. Gazette 42: 265-291, 1906.

9. Davenport, C.B., Germination and solution, Experimental
      morphology, 363.

10. Davenport, E., Variation, correlation, and selection,
      Principles of breeding.

11. Detmer, W. and Moor, S.A., Movement and growth, Practical
      plant physiology.

12. DeVries, H., Hastening of germination, Plant breeding 262.

13. DeVries, H., Variation and correlation, Species and
      varieties.

63

BIBLIOGRAPHY (continued)

14. Duvel, J.W., Vitality and germination of seeds, U.S.Dept.
    of Ag. Bureau of Plant Ind. Bul. 58, May 28, 1904.

15. East, E.M., Selection of seeds in potato growing, U. of I.
    Ag.Exp.Sta. Circular 81.

16. Goodale, G.L., Germination and growth, Physiological
    botany.

17. Harter, L.L., Variability of wheat varieties in resistance
    to toxic salts, U.S.Dept. of Ag., Bureau of Plant
    Ind. Bul. 79.

18. Hodge, C., Natural selection, What is Darwinism?

19. Hume, A.N., The testing of corn for seed, U. of I. Ag.
    Exp. Sta., Bul. 96.

20. Jeffery, J.A., Corn, selection, storing, and curing and
    testing for seeds, Michigan Ag. Exp. Sta. Circular 3.

21. Jordan, D.S., and Kellogg, V.L., Variation and mutation,
    Evolution and animal life.

22. Jordan, D.S., Variation of plant evolution.

23. Jost, L., (translated by Gibson, R.J.H.) Influence of
    external condition in growth, Plant physiology.

24. Kellogg, C., Variation and correlation, Darwinism To-day.

25. Lee, F.S., Stimuli and their actions, General Physiology
    Vernon.

26. Macdougal, D.T., Resistance and acclimatization of seed
    to heat, Practical Text-book of Plant Physiology,
    92-93.

64

BIBLIOGRAPHY (continued)

27. Mitchell, P.C., Weismann theory of the germplasm, The
    Biological Problem.
28. Moorhouse, L.A., Alfalfa seed in Oklahoma, Oklahoma Ag.
    Exp. Sta. Bul.83.
29. Pammel, L.H., and King, C.M., Results of seed investi-
    gation, Iowa Ag. Exp. Sta. Bul. 115.
30. Parker,W.N. and Rownfeldt, H., Variation and heredity,
    The Germ plasm.
31. Percival, J., Speed of germination, Ag. Botany, Theoreti-
    cal and Practical.
32. Pfeffer, W., (translated by Ewart, E.M.) Germination,
    Physiology of Plants.
33. Pfeffer, W., Germination, The Physiology of Plants.
34. Poulton, E.B., Schonland, S., and Shipley, A.E., Heredity,
    Weismann on Heredity.
35. Reid, G.A., Correlation, Heredity, and Variation, The
    Principles of Heredity.
36. Rietz, H.L. and Smith, L.H., The measurement of correla-
    tion, U. of I. Ag. Exp. Sta. Bul. 148.
37. Romanes, G.J., Heredity, An examination of Weismann.
38. Schimper, A.F.W., Germination, dependence on temperature,
    Plant Geography upon a Physiological Basis, 46.
39. Seward, A.C., Variation, Darwin and Modern Science.
40. Shamel, A.D., Improvement of Tobacco by Breeding and
    Selection, Year Book, Dept. of Ag., 1904, pp.440-452.

65

BIBLIOGRAPHY (continued)

41. Strasburger, E., Nole, F., Schenck, H., and Schimper, A.F.W.
        (translated by Porter, H.C.) Germination, Textbook
        of Botany.

42. Strasburger, W. (translated by Hillhouse, W.) Fermenta-
        tive changes in germination, Handbook of Practical
        Botany, 422.

43. Stone, G.E., Influence of chemical solution upon the
        germination of seeds, Rept.Mass.Ag.Exp.Sta. 13,
        74-83, 1901.

44. Thomson, J.A., Variation and Heredity, Heredity.

45. True, A.C., Stimulants for seeds, U.S.Dept. of Ag.,
        Exp. Sta. Work, Vol.II, No.14.

46. Vernon, H.M., Variation, Heredity, and Correlation,
        Variation in Animal and Plant.

47. Webber, H.J. and Boykin, E.B., Advantages of Planting
        heavy cotton seed., U.S.Dept. of Ag., Farmers' Bul.
        285.

48. Weismann, A., Heredity, Essays upon Heredity.

49. Westermaier, M. and Schneider, A., The function of roots,
        A compendium of General Botany.

50. Wiancho, A.T. and Christie, G.I., The selection, preser-
        vation, and preparation of seed corn, Purdue Univ.
        Ag. Exp. Sta. Circular 2.

51. Woods, C.D., Seed Inspection, Maine Ag.Exp.Sta. (Of-
        ficial Inspection 17.)

CPSIA information can be obtained
at www.ICGtesting.com
Printed in the USA
BVHW040104220119
538278BV00008B/140/P